Science Testers

BIOLOGY OF MAN

Mike Lyth

HART-DAVIS EDUCATIONAL

Contents

Introduction

Science Testers arose out of a personal conviction that every good science lesson starts by posing a set of questions and goes on to provide the experiences, activities, and information, which will supply some of the answers. Over the years I had worked out a long list of questions which I felt represented the key facts, concepts, and patterns, in a number of integrated science topics. I was thinking of pupils aged between 13 and 16 and I was heavily influenced with regard to both philosophy and content by the Nuffield Secondary Science Scheme.* My questions were intended to be 'bona fide' in the sense that they were the sort of questions which an average intelligent person might wish to ask, and not those which might only be relevant to the taking of an examination. They were not designed primarily for testing; rather they were intended to provoke discussion, controversy, and enquiry; and to form the basis for individual study and for project work. Ultimately I wrote the questions down in sets, choosing the coded-answer format because it requires minimal written response, allows fast and objective comparison of answers; because it provides, in the distractors, a series of starting points. Many of the distractors came from 'daft answers' which had turned out to be far less 'daft' than they sounded!

Once written and collated, the questions were useful for revision and reinforcement, and were sometimes quite good for testing, though using them for that purpose was not a step to be taken lightly. Answering questions in a relaxed classroom atmosphere is one thing, answering them in a formal test situation quite another. Formal tests and examinations have serious effects on the personalities, not to say on the future life styles, of people taking them, so the questions and items making up the tests must be taken seriously. Many of the Science Tester questions were taken seriously in this way, being tested in examination conditions and refined both, by myself and by other teachers. With a bit of work more could be made more effective as test items, but then they would probably be less interesting and less effective as discussion starters! However, the Item Writers Handbook aims to provide help for teachers wishing to do that with them, as well as help in writing items from scratch. It also shows how items can be tested statistically. I hope it turns out to be useful, and that both it and the Testers are enjoyable and interesting.

*Nuffield Secondary Science.
Organizer: Mrs. Hilda Misselbrook.
Materials published for the Nuffield Foundation by Longman Group Ltd.

1. Physical Activity

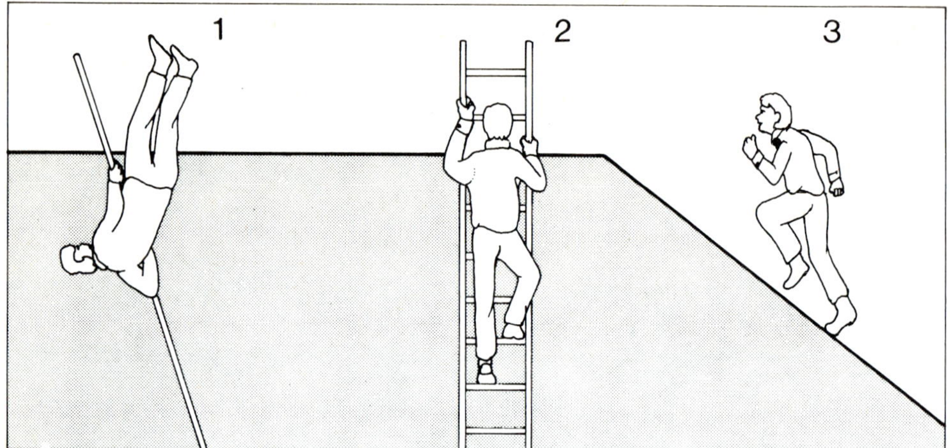

1 All the boys weigh the same. Each gets onto the platform in his own way. Which of the following statements is true?

A Boy **1** does the most work
B Boy **1** does the least work
C Boy **2** does the most work
D Boy **3** does the most work
E All do roughly the same amount of work

2 Susan has just finished running upstairs. The height of the stairs is 10 metres. She weighs 500 newtons. How much energy has she converted into uphill (stored) energy? In the right units it is:

A 5 **B** 10 **C** 50 **D** 500 **E** 5000

3 If Susan runs upstairs in 20 seconds, how much energy does she convert per second? In the right units is it:

A 250 **B** 500 **C** 1000 **D** 2500 **E** 100 000

4 The right units to use for energy converted are:

A joules **B** watts **C** newtons
D metres/second **E** newtons/metre

5 Energy converted per second is called:

A force **B** work **C** heat
D power **E** momentum

6 The people in the picture are using a 'cycle ergometer' to test their ability to convert energy. The sandbag **S** weighs 20 newtons. Before the boy starts pedalling the forcemeter **F** will read:

A zero **B** 10 N **C** 20 N **D** 30 N **E** 40 N

7 When the boy starts pedalling the forcemeter reading will probably fall by about 5 newtons. If it falls by exactly 5 newtons the friction force he will be working against will be:

A 5 N **B** 10 N **C** 15 N **D** 20 N **E** 25 N

8 In an experiment like the one in questions 6 and 7 several people worked on the ergometer. The amount of air that each of them breathed out was collected and measured. What conclusion do you think they came to? Was it:

A the faster you convert energy the more air you need
B the more energy you convert, the more air you need
C the slower you convert energy the more air you need
D you always breathe out the same amount in the same time
E everybody needs the same amount of air to do the same job

ACTIVITY						
AIR BREATHED PER MINUTE	**6**	**7**	**8**	**14**	**43**	**65**
ENERGY CONVERTED PER MINUTE	**5**	**6**	**7.5**	**13.5**	**41**	**61**

9 The table shows the amount of air breathed (in litres) and the energy converted (in kilojoules) whilst various things are being done. From the information in the table, which of the following statements is/are true?

Statement **1** The more vigorous the activity the more energy you need per second
Statement **2** The more vigorous the activity the more air you need per second
Statement **3** You always get roughly the same amount of energy from a litre of air.

Choose your answer letter using the code below:

Code Choose **A** if **1 2 3** are all true
B if **1 2** only are true
C if **2 3** only are true
D if **1** only is true
E if **3** only is true

10 In order to keep on converting energy, your body must have one of these gases. Which one?

A nitrogen **B** oxygen **C** hydrogen
D helium **E** carbon dioxide

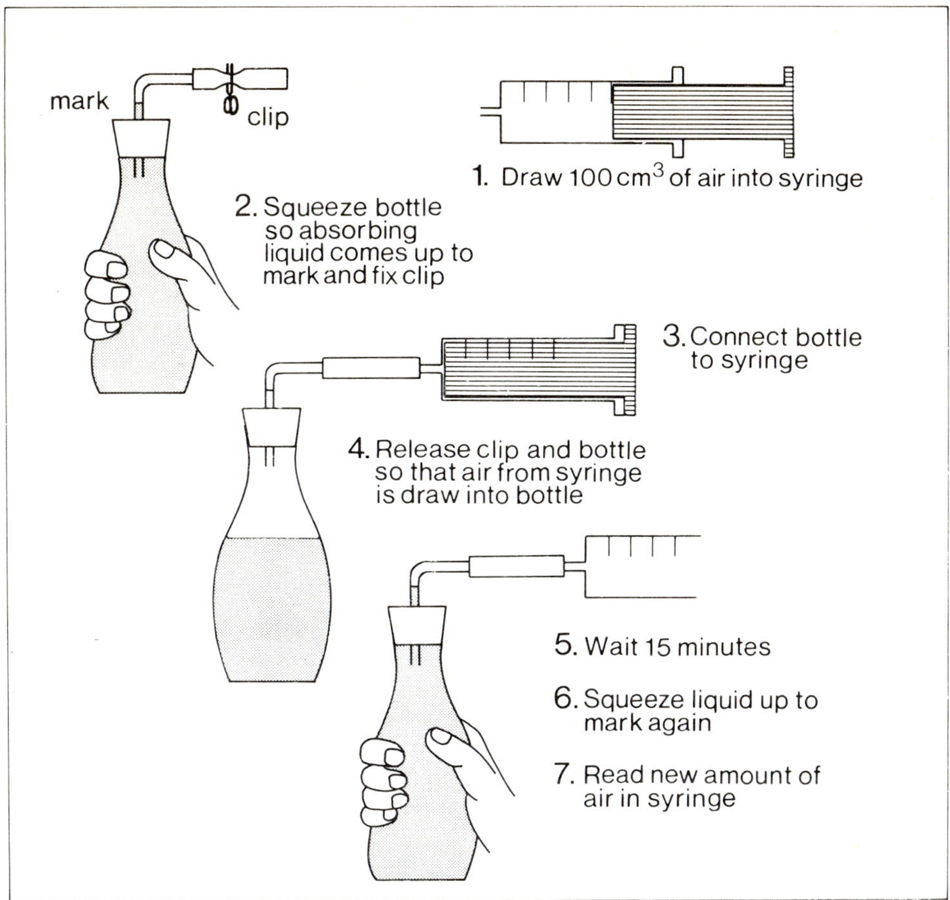

The diagram shows a way of finding out how much carbon dioxide (CO_2) there is in a sample of air. (The method uses a liquid which absorbs CO_2.) The next *six* questions are about this experiment.

11 Why was the bottle squeezed at the beginning? Was it to:

A spread the liquid over the sides of the bottle
B force air out of the bottle
C mix up the liquid
D make sure there was enough liquid in the bottle
E warm up the liquid

12 Why was the bottle left for 15 minutes (Stage 5)? Was it to:

A let the liquid drain off the sides of the bottle
B let the liquid cool down after squeezing
C give the atmosphere time to push the plunger in
D give the liquid time to absorb the gas
E give the experimenters time to think

13 To find the amount of carbon dioxide, what liquid would you expect to be in the bottle? Would it be:

A potassium hydroxide solution
B potassium permanganate solution
C pyrogallol solution
D sulphuric acid
E water

14 If you drew 100 cm^3 of air into the syringe to start with (taking the air from the lab.) what would you expect the syringe to look like at the end? Choose one of the diagrams labelled A to E.

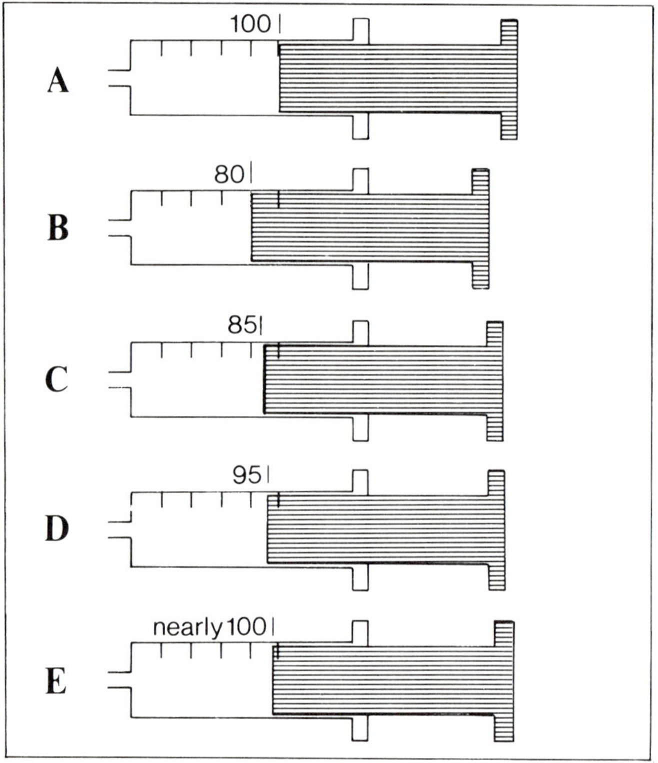

15 If you had filled the syringe with air from your lungs (by breathing into it) which of the diagrams would it look like at the end?

16 If you wanted to find out how much *oxygen* there was in the air in the lab., what liquid would you put in the bottle? Choose one of the list in question 13.

17 Oxygen is absorbed into the bloodstream through the walls of very fine blood vessels—capillaries. Where does this go on?

A around the stomach walls
B through the walls of the intestine
C around the lung bronchi
D in the windpipe
E around the lung air sacs

18 Which of the following three things is/are used up during physical activity? Use the code below to choose your answer letter.

	1 food	**2** oxygen	**3** carbon dioxide
A	used up	used up	used up
B	used up	used up	not used up
C	not used up	used up	used up
D	used up	not used up	not used up
E	not used up	not used up	used up

19 Four of the following are lost when you sweat. Which one is *not* lost?

A water **B** minerals **C** skin bacteria
D weight **E** heat

20 Which of the following helps to lower someone's body temperature?

A shivering **B** sweating **C** hair standing on end
D less blood in skin capillaries **E** drinking tea

21 Before dissecting a sheep's lung some children were asked to say what they thought lungs were like. Which do you think is the *best* description?

A 'lungs are like balloons made of thin elastic stuff—they fill with air when you breath in'
B 'lungs are like balloons made of thin elastic stuff and filled with blood to soak up oxygen'
C 'lungs are like balloons made up of strips of gristle with muscles in between'
D 'lungs are like sponges full of tiny air bags which fill with air when you breathe in'
E 'lungs are like sponges full of tiny bags holding blood to soak up oxygen'

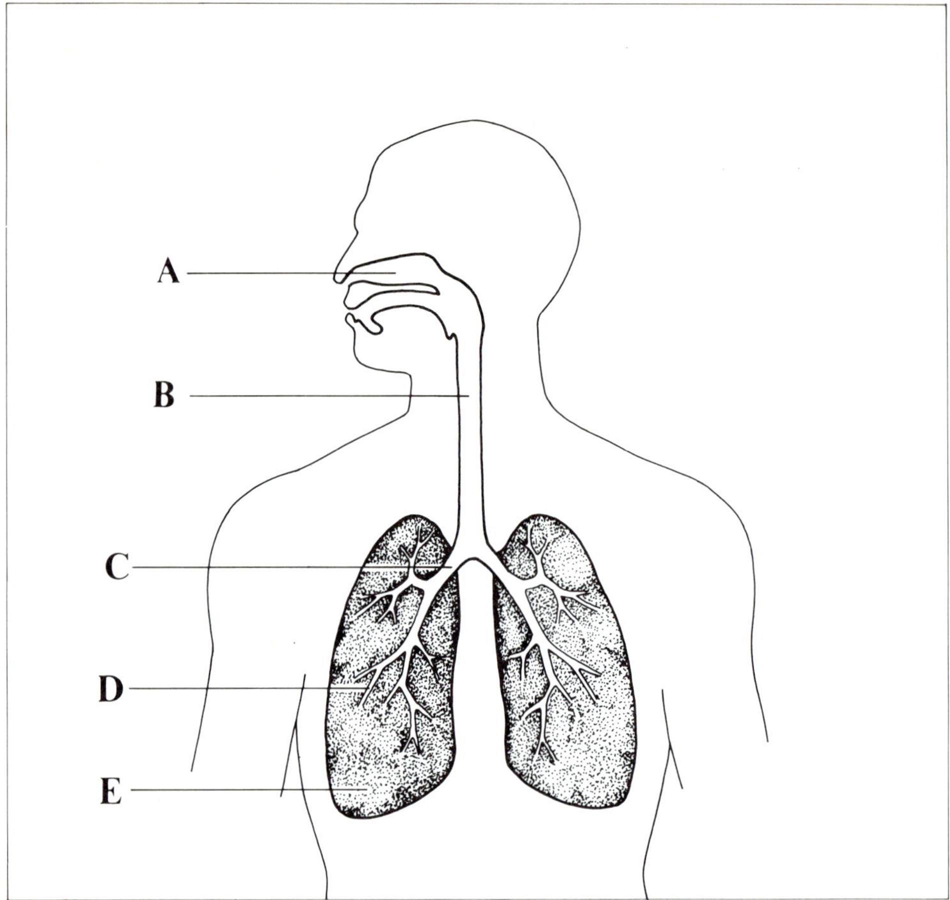

Choose one of the parts marked A to E in the diagram to fit the labels in the next three questions

22 'the windpipe'

23 'a bronchus'

24 'many alveoli here'

25 The tubes to the lungs are lined with tiny hair-like things called cilia. What do they do?

A stop food going down the wrong way
B sweep mucus and dust up and out
C absorb oxygen from the air
D keep the air moving through the tubes
E stop the inside of the tubes drying up

26 Roughly how much air goes into and out of your lungs in a normal breath? In litres is it:

A $\frac{1}{2}$ to 1 **B** 2 to 3 **C** 4 to 5 **D** 6 to 7 **E** 8 to 9

27 About how much air in litres would your lungs hold if you breathed in really hard? Choose one of the numbers above.

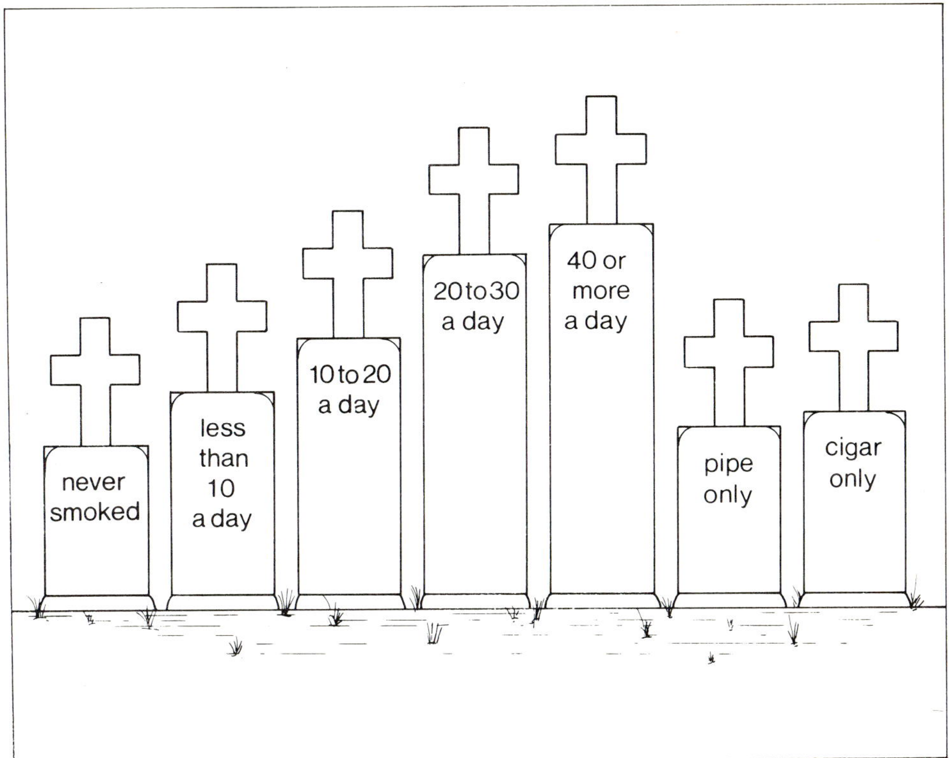

28 The diagram shows some statistics from research into smoking and health. Going on the results in the diagram, which of the following statements is/are true?

1 The less you smoke the more likely you are to survive
2 Smoking causes lung cancer
3 Pipe smoking is harmless

Choose your answer letter using the code below:

Code	Choose					
	A if	**1**	**2**	**3**	are all true
	B if	**1**	**2**		only are true
	C if		**2**	**3**	only are true
	D if	**1**			only is true
	E if			**3**	only is true

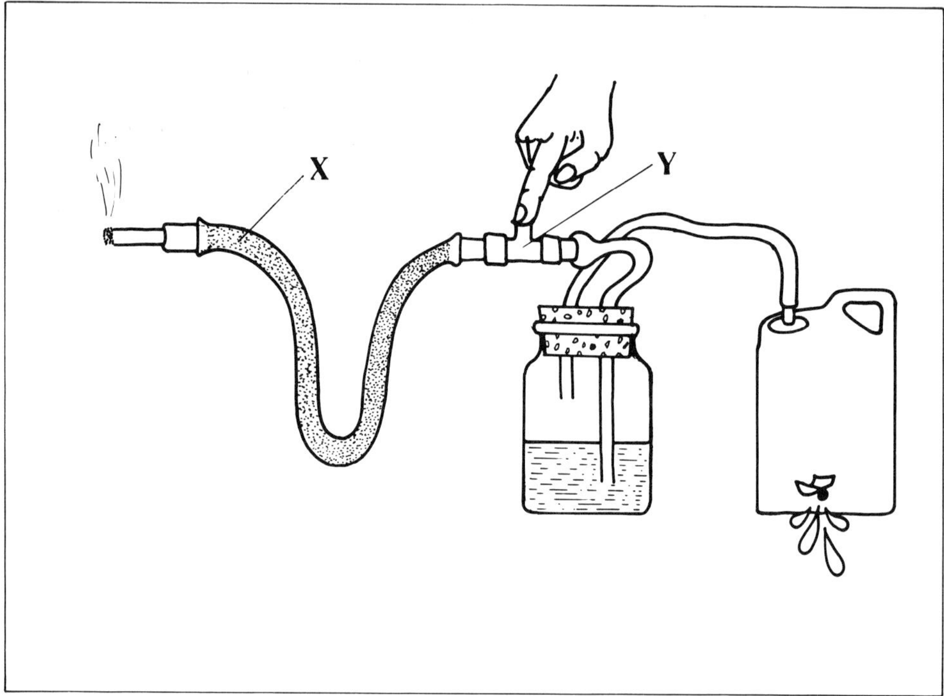

29 The diagram shows a home made 'smoking machine' for testing cigarettes. Which of the following labels below is the best one for part **X**?

A lets air into the apparatus
B makes the cigarette puff
C collects solids and tar
D stops smoke coming through too quickly
E makes the air come through

30 Which of the labels above is the best for part **Y** in the diagram?

31 Which of the following muscles helps to straighten the arm?

A biceps **B** triceps **C** hamstring **D** deltoid **E** trapezius

32 Which of the muscles in the list helps to bend the knee?

33 Where is the cardiac muscle found? Is it in the:

A lungs **B** legs **C** heart **D** stomach **E** back

34 Which of the following parts of the skeleton protects the heart and lungs?

A ribcage
B skull
C backbone
D pelvic girdle
E shoulder blades

35 Which of the parts in the list above protects the nerve cord?

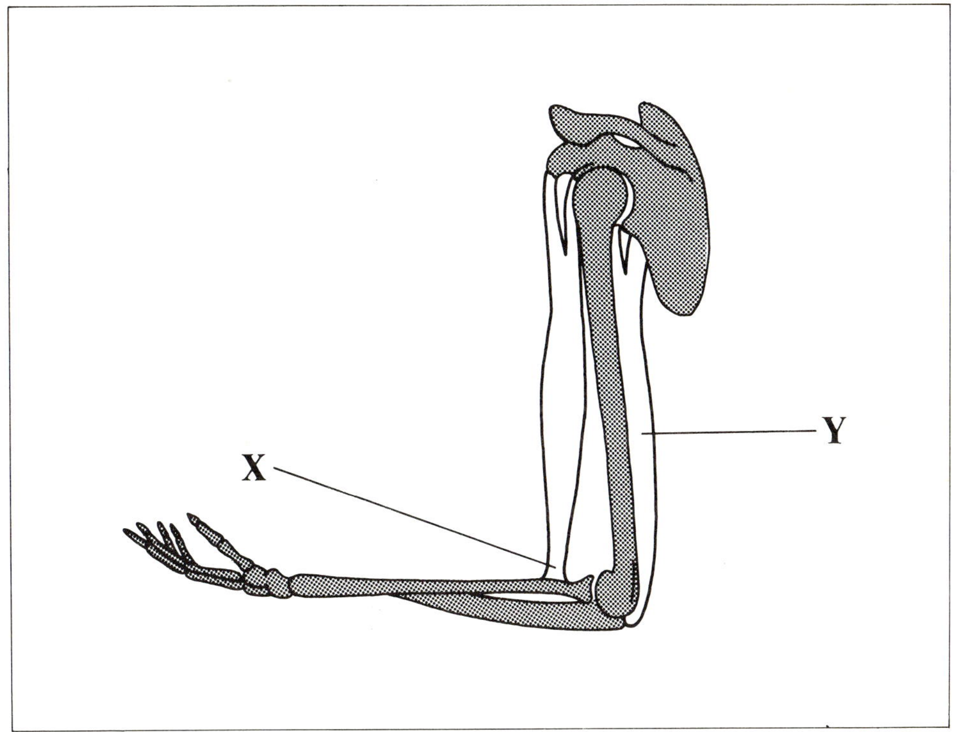

36 Which of the labels below should be used for part **X** in the diagram?

A biceps muscle
B triceps muscle
C tendon
D bone marrow
E cartilage

37 Which of the labels in the list should be used for part **Y**?

38 Which of the list best fits the definition: 'shock absorbing material found on the ends of bones'?

39 Which of the list best fits the definition: 'where red cells are made in adults'?

40 Which of these statements about muscles is/are true?

1 some muscles push on bones, others pull on them.
2 muscles always work in pairs or sets
3 muscles never quite stop moving.

Choose your answer letter using the code below:

Code	Choose						
		A	if	**1**	**2**	**3**	are all true
		B	if	**1**	**2**		only are true
		C	if		**2**	**3**	only are true
		D	if	**1**			only is true
		E	if			**3**	only is true

2. Food and Transport

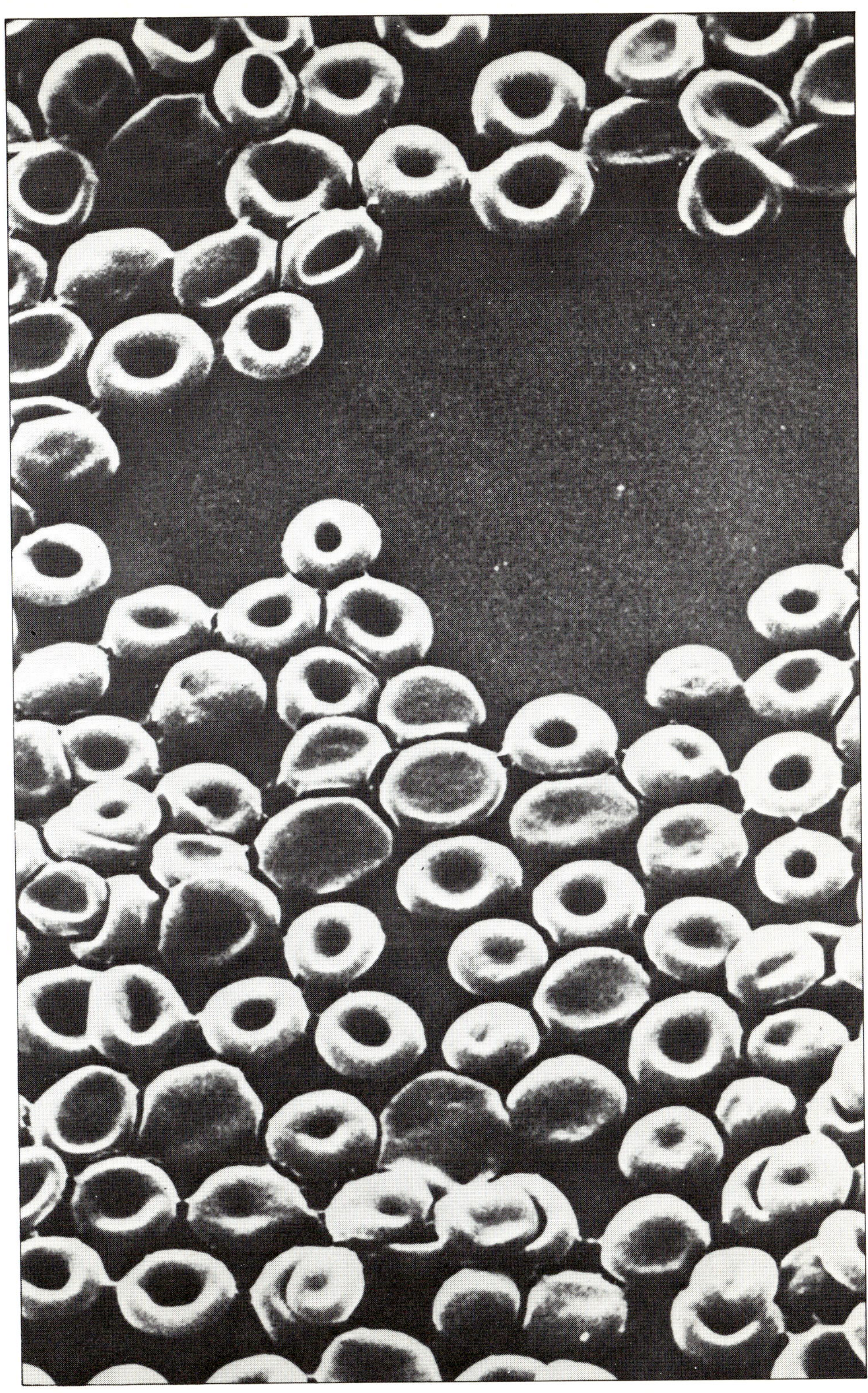

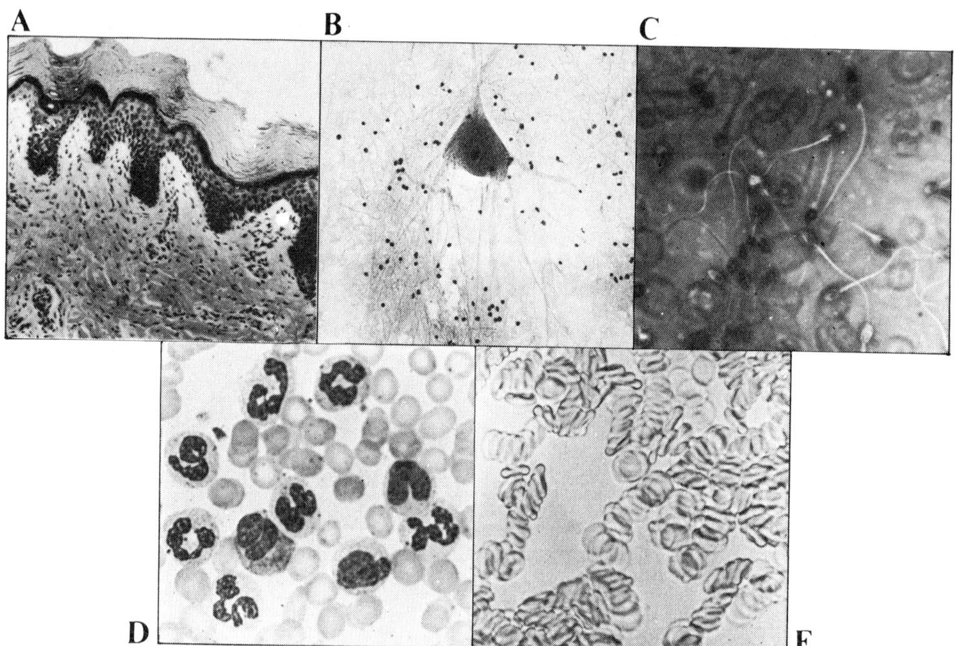

1 Which of the pictures marked **A** to **E** above, shows red blood cells only?

2 Which of the pictures shows white blood cells?

Here is a list of things found in blood. Use it to answer the next *four* questions:

A red cells
B white cells
C salt
D insulin hormone
E adrenalin hormone

3 Which of the list carries oxygen through the bloodstream?

4 Which helps destroy bacteria?

5 Which carries messages about blood sugar?

6 Which alerts the body for vigorous action ('fight or flight')?

7 Which one of the following statements about the heart is true?

A it takes blood from veins
B it takes blood from arteries
C it has two chambers
D it beats at a steady rate all the time
E it beats at the same rate for everybody

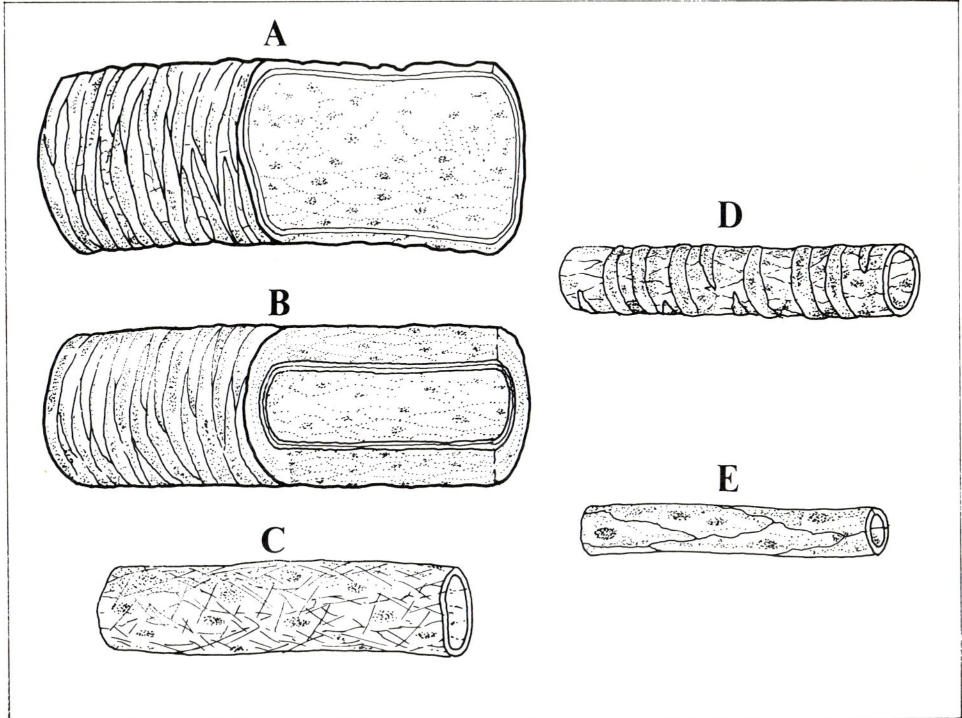

8 Which of the blood vessels labelled **A** to **E** in the diagram would you say was an artery?

9 *Four* of the following five statements about white cells are true. Which of them is *false*?

A they surround and destroy bacteria
B they carry oxygen through the bloodstream
C they produce antibodies to neutralize strange substances
D they remove damaged or dead cells
E they are made in the spleen and lymph glands

10 The liver does four of the following things. Which one does it *not* do?

A stores fat
B makes blood proteins
C stores vitamins
D makes red cells in adults
E makes bile

Use this list of foods for the next four questions:

A milk
B potato crisps
C cod liver oil
D oranges
E lettuce

Which of the foods in the list would be best to eat for:

11 vitamin C

12 vitamin A

13 calcium

14 fat

Use this list of 'things that your food is useful for' to answer the next four questions:

A supplying energy
B keeping you fit and healthy
C making strong bones and teeth
D protecting you from the cold
E helping your body grow and repair itself

Which of the list is calcium most useful for?

Which way do vitamins help?

17 Which is sugar most important for?

18 How do proteins help?

19 All the things below are important in our diet, but one of them passes straight through our bodies. Which one?

A cellulose **B** protein **C** starch
D vitamins **E** iron

20 Which of the nutrients below give you energy?

A proteins **B** fats **C** starches
D sugars **E** all of these

Use the table of food tests below and the code under it to answer the next six questions.

TEST	COLOUR AT START	COLOUR AT END	SHOWS PRESENCE OF
1 Add ninhydrin solution and warm	Pale yellow	Purple	Protein
2 Add iodine solution	Brown	Blue Black	Starch
3 Add benedict's solution and boil	Pale blue	Golden yellow	Glucose sugar

Choose letter
A if tests **1 2 3** would work
B if tests **1 2** only would work
C if tests **2 3** only would work
D if test **1** only would work
E if test **3** only would work

Which test or tests would work with:

21 Pure egg white?

22 Milk?

23 Plain sponge cake?

24 Sausages?

25 Jam and bread?

26 A new food which is claimed 'to supply energy fast'.

27 If you want to test for double sugar you begin by warming your sample with dilute hydrochloric acid (HCl). Why? Is it to:

A split the double sugar up
B dissolve the sugar
C produce carbon dioxide gas
D get rid of protein and fat
E decolourize the food

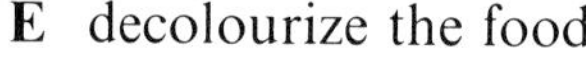

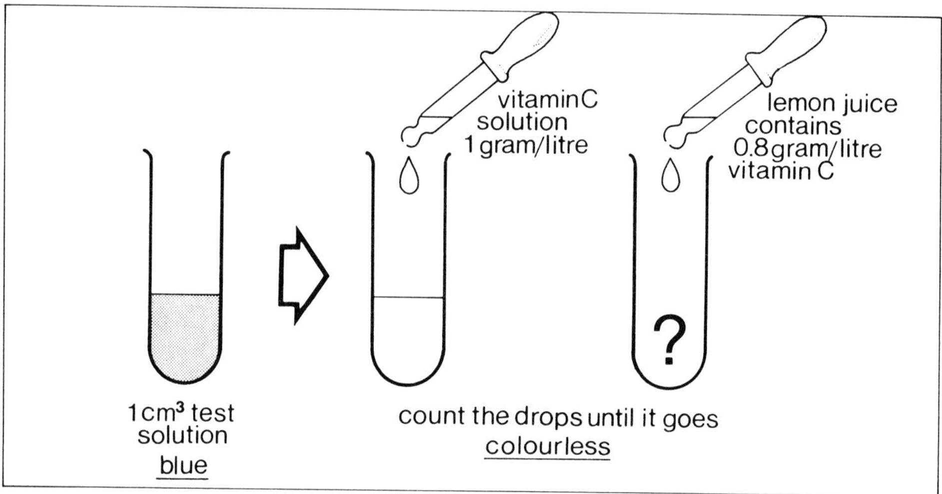

The drawing shows a way of testing how much vitamin C there is in different foods. In one experiment it took 15 drops of pure vitamin C solution to decolourize the test solution.

Use the following list to answer the next three questions:

A only 1 drop
B about 15 drops
C about 20 drops
D about 25 drops
E at least 50 drops

28 About how many drops of lemon juice would be needed to do the same thing?

29 Suppose pure water was dropped into the test solution. How many drops would be needed to change the colour then?

30 What if the lemon juice had been 'cooked' (heated for a few minutes) before being dropped in. How many drops would be needed then?

31 Here are the names of some parts of the human digestive system:
1 mouth **2** large intestine **3** stomach **4** small intestine

In what order does the food pass through the four parts? Is it:

A 1–2–3–4 **B** 1–3–2–4 **C** 1–4–3–2 **D** 1–4–2–3 **E** 1–3–4–2

Choose one of the parts of the digestive system marked **A** to **E** on the diagram to answer the next four questions.

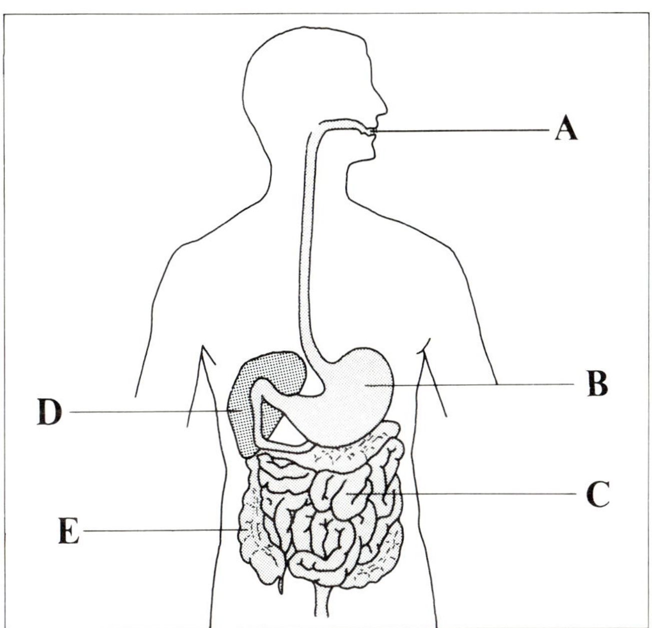

32 Which of the parts is the liver?

33 Where does this happen: 'the food is mixed with digestive fluids and almost all of the valuable nutrients are removed'?

34 In which part does this happen: 'the food is mixed with acid and digestive fluids'?

35 In which part does this happen: 'the food is broken up, moistened and rolled into a ball'?

36 How does food get through the tube joining part A to part B in the diagram? Is it mainly:

A pushed along by food behind it
B pulled down by its own weight
C washed down by digestive juices
D squeezed down by muscles
E swept down by little hairs in the tube

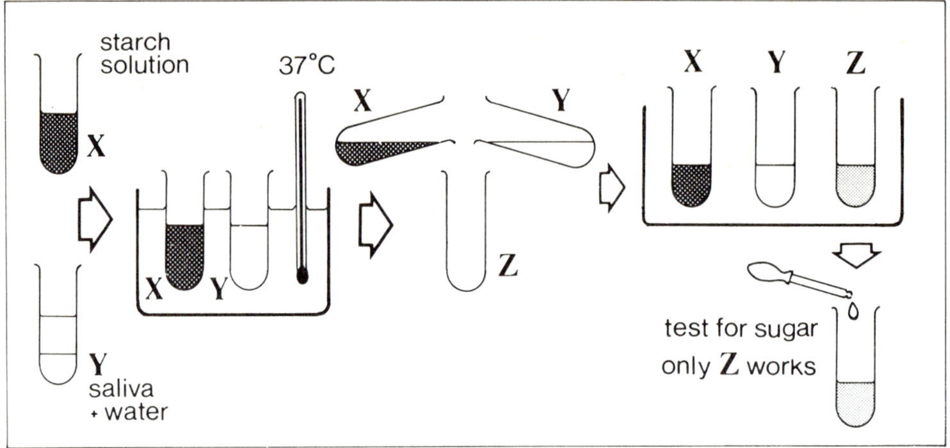

The diagram shows a way of finding out what happens to starch when it is mixed with saliva in our mouths. Each of the questions about the experiment has three answers numbered 1, 2, 3.

Choose your answer letter using the code below:

Code	Choose					
	A if	**1**	**2**	**3**	are all true
	B if	**1**	**2**		only are true
	C if		**2**	**3**	only are true
	D if	**1**			only is true
	E if			**3**	only is true

37 At the end, test tube **Z** contained:
1 starch **2** saliva **3** water

38 All three test tubes had to be tested because:
1 the starch could have contained sugar to start with
2 the saliva could have contained sugar to start with
3 the test tubes could have been dirty

39 The experiment tells us that:
1 saliva can change starch into sugar
2 wetting starch changes it into sugar
3 heating saliva changes it into sugar

40 Tubes A and B were put in a water bath:
1 to dissolve the starch
2 to make the starch into sugar
3 to heat them to mouth temperature

3. Microbes

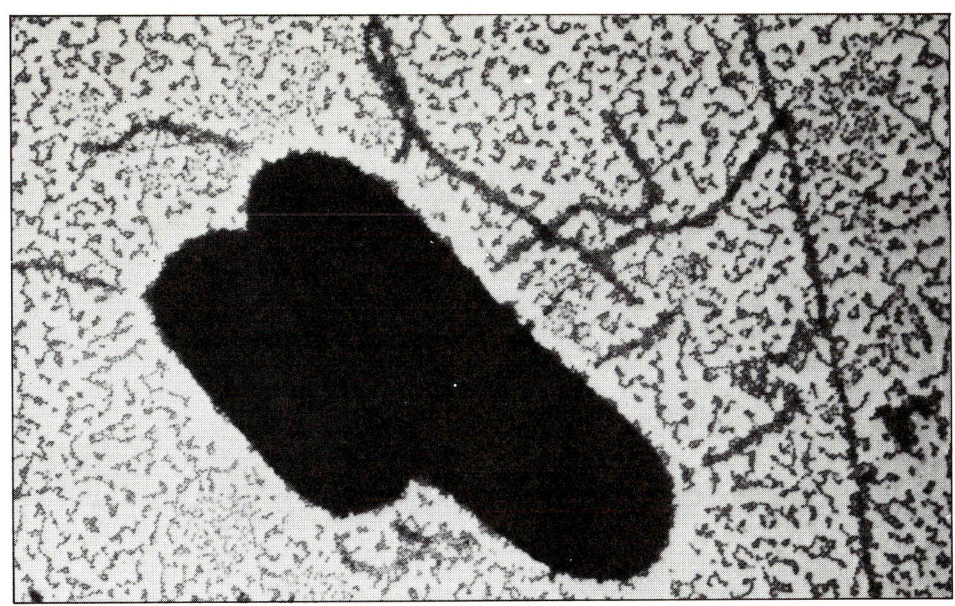

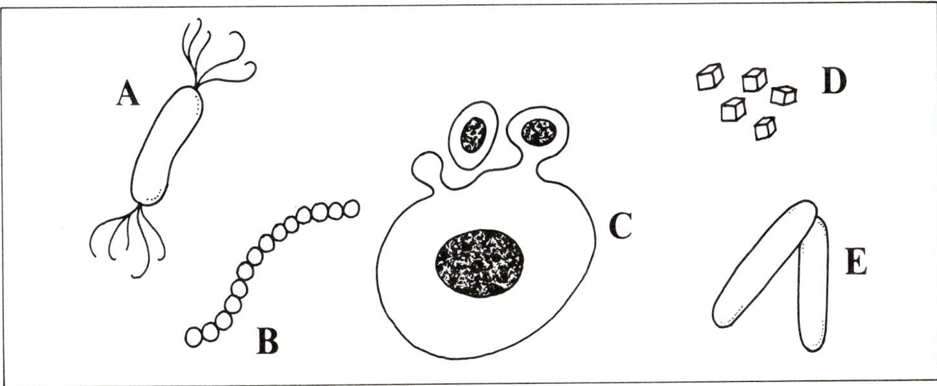

1 Which of the pictures of microbes marked **A** to **E** would you say shows a *virus*?

2 Which of the pictures would you say shows a *yeast*?

3 Suppose you dug up a teaspoonful of soil from the garden. How many bacteria would there be in it? Would the number be roughly the same as the number of people:

A in your family
B in a cinema
C in a football crowd
D living in a big city
E in the whole world

4 Which of these three statements about bacteria is/are true?

1 Some bacteria are helpful to humans
2 If your skin is clean it should be free of bacteria
3 All bacteria need oxygen to survive.

Choose your answer letter using the code below:

Code	Choose					
	A	if	**1**	**2**	**3**	are all true
	B	if	**1**	**2**		only are true
	C	if		**2**	**3**	only are true
	D	if	**1**			only is true
	E	if			**3**	only is true

5 Which of these statements about bacteria is/are true?
Use the code in question **4** to choose your answer letter.

1 Refrigeration kills bacteria
2 Most bacteria do well in bright sunlight
3 Most bacteria end up by destroying themselves

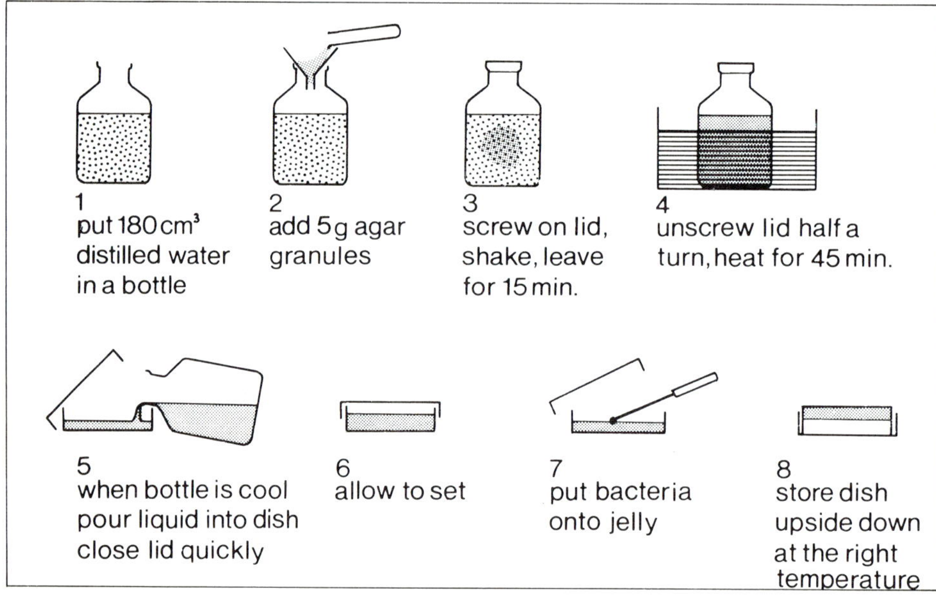

The diagram shows a method of growing (culturing) bacteria. Listed below are reasons for doing some of the things in the diagram. Use the list to answer the next three questions.

A to let the bacteria in
B to keep the bacteria out
C to let steam and air out
D to stop the top sticking
E to stop condensed water dripping on the agar

6 Why was the bottle's lid unscrewed before 'cooking' it?

7 Why was the lid of the dish closed *quickly* in picture 5?

8 Why was the dish stored upside down in picture 6?

9 What should the 'right' temperature be in picture 8?

A 25°C **B** 37°C **C** 50°C **D** 63°C **E** 100°C

10 Suppose you started with only 20 bacteria in the dish, how many would you expect to find $1\frac{1}{2}$ hours later?

A 20–30 **B** 30–100 **C** 100–1000
D 1000–10 000 **E** millions

Here is a list of methods of preserving food:

A curing (salting) **B** canning **C** quick freezing
D pasteurizing **E** pickling

Use the list to answer the next three questions.

11 Which of the methods completely kills the microbes in the food?

12 Which makes the food too acid for microbes to grow?

13 Which takes away the water that they need to grow?

14 Why does cooked food go bad? Is it because:

A some of it changes into bacteria
B the heat used in cooking causes decay
C bacteria in the air feed on it
D oxygen in the air causes decay
E flies lay their eggs in it

To test the freshness of samples of milk you can add a blue dye to it. If there are bacteria living in the milk, the colour of the dye changes from blue to pink. In an actual test some samples of milk were made ready like this:

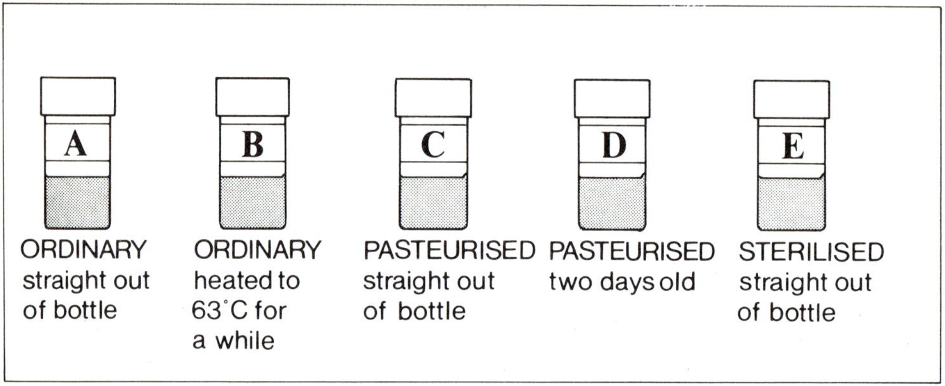

The dye was added and the samples were kept at blood heat.

15 Which tube would go pink first?

16 Which tube would go pink second?

17 Which tube would go pink last?

18 Bacteria play a part in four of the following. Which one are they *not* involved in?

A wine making **B** cheese making **C** butter making
D treatment of sewage **E** improving the soil

19 What kind of microbes are involved in bread making? Are they:

A bacteria **B** viruses **C** fungi **D** algae **E** protozoans

20 Why does bread *not* go on 'rising' *all* the time it is being cooked? Is it mainly because:

A the microbes die in the heat of the oven
B all the microbes are used up after a while
C the gas the microbes make escapes from the bread
D the sugar in the dough poisons the microbes
E the bread gets too hard to swell up

Here is a list of things which cause disease:

A a bacterium **B** a fungus **C** a virus
D shortage of a vitamin **E** tiny single celled animal (protozoan)

For each of the diseases in the next five questions, choose from the list the thing that causes it.

21 malaria

22 rickets

23 athlete's foot

24 measles

25 typhoid

Choose a word from the list below to fit each of the descriptions in the next three questions:

A antiseptic **B** antibiotic **C** antitoxin
D antigen **E** antibody

26 substances formed by the blood which cause poisons to lose their effect.

27 substances formed by the blood which make you immune to a disease.

28 substances which kill microbes or stop them growing, e.g. penicillin.

Choose one of this list of things found in blood to answer the next two questions:

A red cells **B** white cells **C** platelets
D antibodies **E** hormones

29 Which of the list will surround and destroy invading bacteria?

30 Which are particularly important in blood clotting?

Read the notes on blood grouping opposite, then try to answer the next four questions.

31 Suppose you had a patient with type 3 blood. Which type would it be safe to give him?

A type 1 **B** type 2 **C** type 3 **D** type 4 **E** all types

32 If a patient can take blood from anyone, what type of blood must he have?

33 Some patients can only take blood of their own type—any other type clogs up straight away. What type do they have?

34 One type of blood can be given to anybody with safety—it has no peculiar red cells. What type is that?

35 What is an allergy? Is it an illness caused by:

A something that is harmless to most people
B narrowing of the heart's blood vessels
C damage to the baby whilst it is still in the womb
D a wrongly adjusted diet
E a parasite growing in the body

36 Which of the list above is probably the cause of *congenital* disease?

37 *Four* of these diseases are spread in little drops of liquid coughed or sprayed out and carried in the air. Which one is *not*?

A common cold **B** smallpox **C** measles
D venereal disease **E** mumps

Data sheet on blood grouping

In a blood transfusion, red cells from a donors blood are put into the patient's blood plasma. The problem is that the patient's plasma may have antibodies in it, which will attack the foreign red cells. Doctors have to be careful to choose the right blood to give the patient.
To show how he does it we can draw diagrams like this...

These stand for the three kinds of red cells that people can have

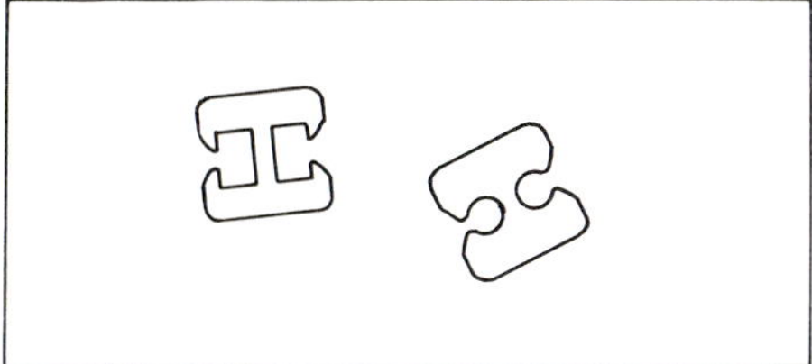

These stand for the antibodies that people can have in their plasma

People can have 4 different types of blood - mixtures of the red cells and antibodies.

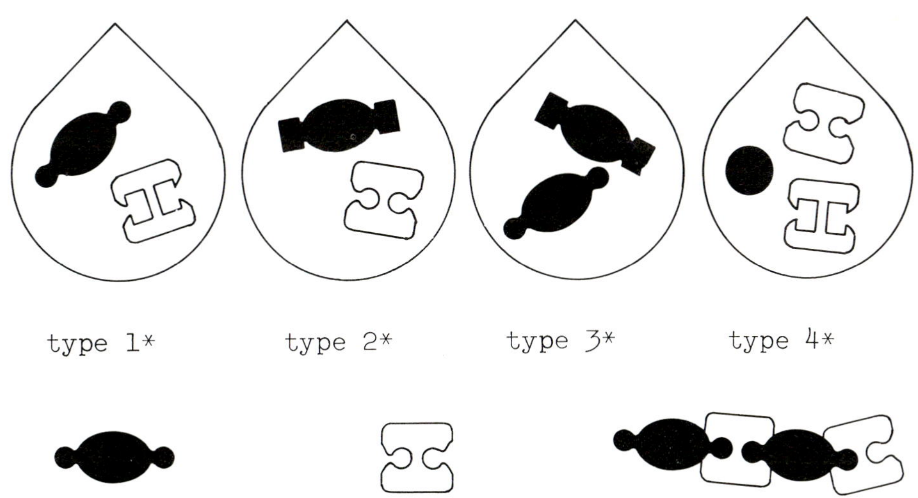

The doctor must not put red cells like this........

into blood with antibodies like this...........

or this will happen- the antibodies join all the red cells together.

So he can't give type 1 blood to someone with type 2. And he can't give type 2 to someone with type 1 or type 4 - perhaps you can see why?

But he can give this type of red cell to anybody..

38 You can be made immune to many diseases by being *vaccinated.* How do vaccines work? Do they:

A poison all the bacteria in your bloodstream
B give you a mild dose of the disease
C bring all the bacteria out in spots and pimples
D cause poisons made by the bacteria to lose their effect
E make the white cells in your blood work harder

39 You can also be injected with *antitoxins.* Which of the list above do they do?

40 Which of the following treatments could be dangerous to someone suffering from shock?

A talking to him constantly
B laying him on his back with his head low
C keeping him cool and in fresh air
D giving him a drink of whisky or brandy
E giving him a drink of water or lukewarm tea

4. Life Cycle

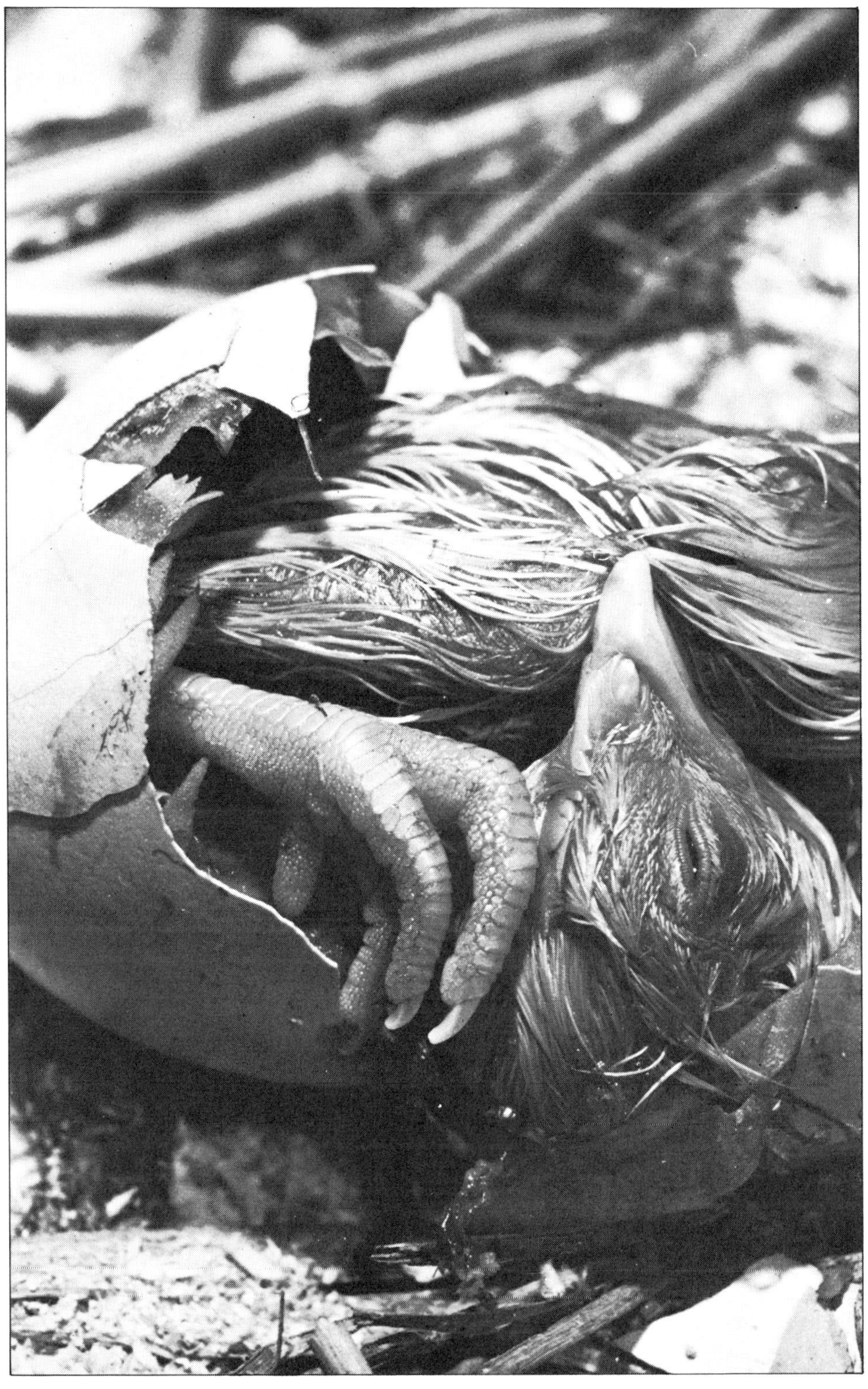

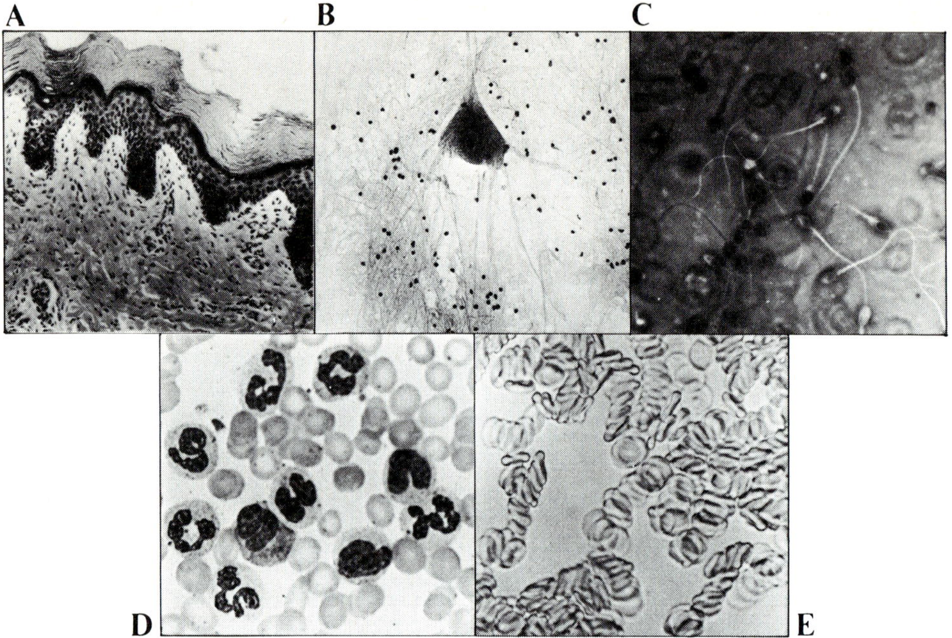

Use these pictures of different kinds of cells found in the human body to answer the next four questions.

1 Which picture shows nerve cells—which have to send out and collect lots of messages?

2 Which picture shows skin cells—which have to form a protective layer over our bodies?

3 Which picture shows male sex cells?

4 Which picture shows cells which have no nucleus?

5 How big is a typical cell? The ones in the pictures are much magnified. For instance, how many could you get side by side between two millimetre marks on your ruler? Would it be:

A 2 or less **B** 10 to 20 **C** 50 to 1000
D 500 to 1000 **E** more than 1000

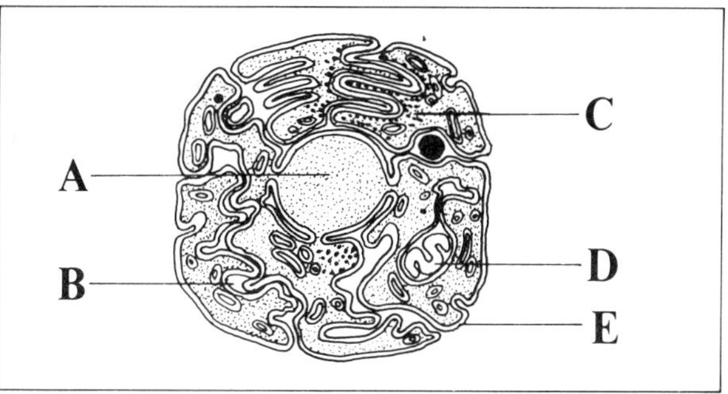

6 The diagram shows an artist's idea of a typical cell. Which of the arrows labelled **A** to **E** points to the *nucleus*?

7 Which of the arrows in the cell diagram points to the *membrane* of the cell?

8 What does the nucleus of the cell do? Does it:

A manufacture new proteins
B control the activities of the cell
C protect the cell from its surroundings
D generate energy for the cell
E split the cell in half when necessary

9 Which of the body's cells carry information that can be used to produce a new person? Is it:

A only the sex cells
B only the blood cells
C only the brain cells
D only the nerve cells
E most of the cells in the body

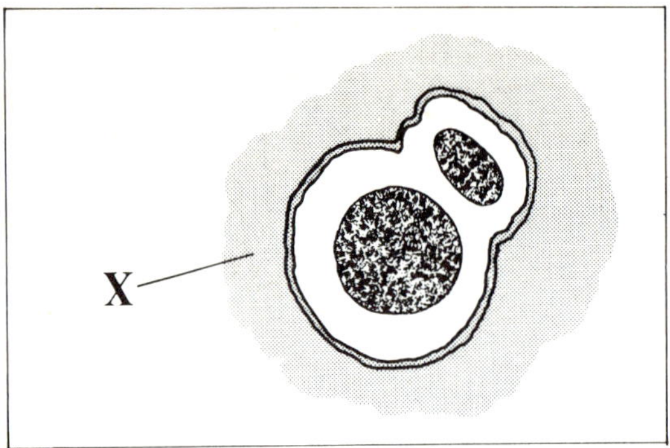

10 In this picture of a human egg cell, what does part **X** do? Does it:

A supply energy to the embryo
B feed the embryo during pregnancy
C control the sex of the embryo
D change into new cells
E only protect the embryo

11 What does *fertilization* mean? Is it:

A the coming together of sperm and egg to form a single cell
B a single egg starting to grow into an embryo
C sperm entering the vagina
D the lining of the womb being renewed
E an egg being planted in the wall of the womb

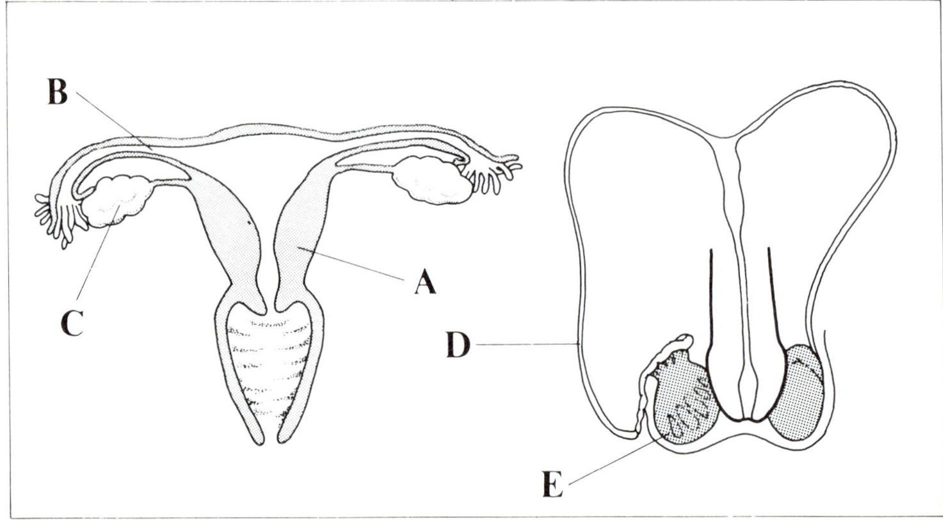

Use the diagrams of the female and male reproductive organs to answer the next five questions.

Which of the parts labelled **A** to **E** in the diagrams is the:

12 egg duct?

13 ovary?

14 sperm duct?

15 testis?

16 womb (uterus)?

17 Where does fertilization take place? Is it in the:

A ovary **B** egg duct **C** vagina **D** testis **E** womb

18 What happens to a human egg if it is not fertilized? Does it:

A embed in the lining of the womb
B go back to the ovary
C pass out of the body with the womb lining
D split open and release blood
E cause the immediate release of another egg

19 What causes identical twins? Do they develop if:

A an *unfertilized* egg splits into two
B a *fertilized* egg splits into two
C one sperm fertilizes two eggs
D two sperms fertilize two eggs
E a sperm splits into two

20 Which of the list above results in fraternal twins?

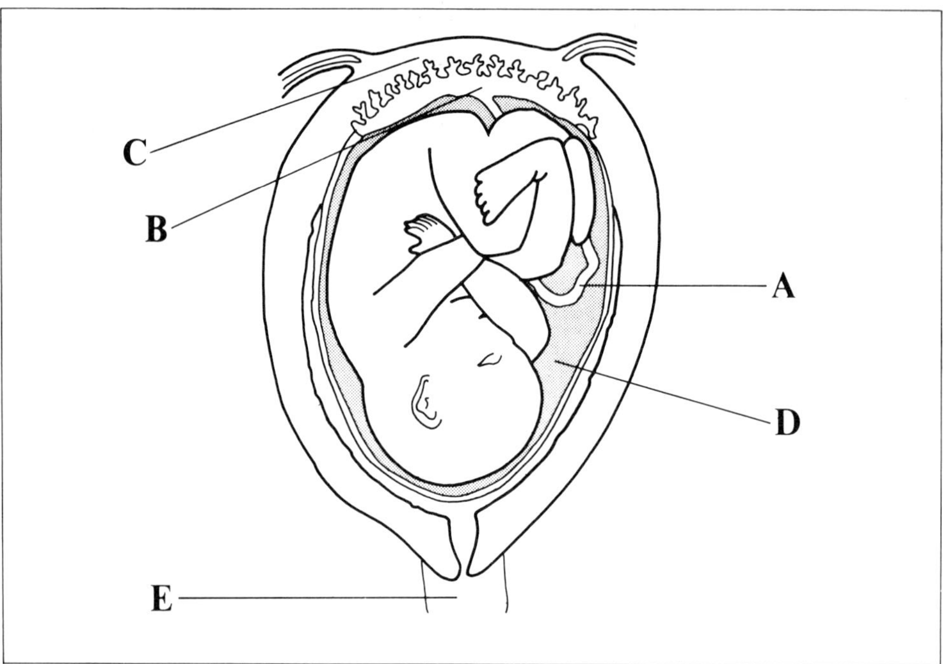

Choose one of the parts labelled **A** to **E** in the diagram to fit each of the descriptions in the next five questions:

21 'one of the places where mother's blood and baby's blood come closest together.'

22 'the birth canal or vagina.'

23 'the cord which carries the baby's blood (umbilical cord)'.

24 'the fluid filled bag (amnion)'.

25 'the organ which feeds the baby and takes away its waste'.

26 How old would a human embryo be to fit this description:
'the embryo has just developed into a rounded little body with head, trunk, and umbilical cord. The beginnings of the mouth, heart, and backbone have just formed?' Would it be:

A 1 week old or less
B 3 weeks old
C 5 weeks old
D 7 weeks old
E 9 weeks old or more

27 Using the list in question **26**, how old would the embryo be to fit this description:
'by the end of the week the embryo is about half an inch long. It's arm and leg buds have just formed':

28 How old would the embryo be to fit this description:
'the embryo has just become a foetus—everything is there that will be there in the newborn baby. Muscles have formed and started exercising'?

29 The fluid-filled bag surrounding the foetus provides *four* of the following. Which does it *not* provide?

A shock absorption
B removal of waste
C heat insulation
D a food supply
E lubrication of the birth canal

30 What two instincts does a *new born* baby need for survival? One is sucking, the other is:

A crying
B grasping
C hearing
D seeing
E smiling

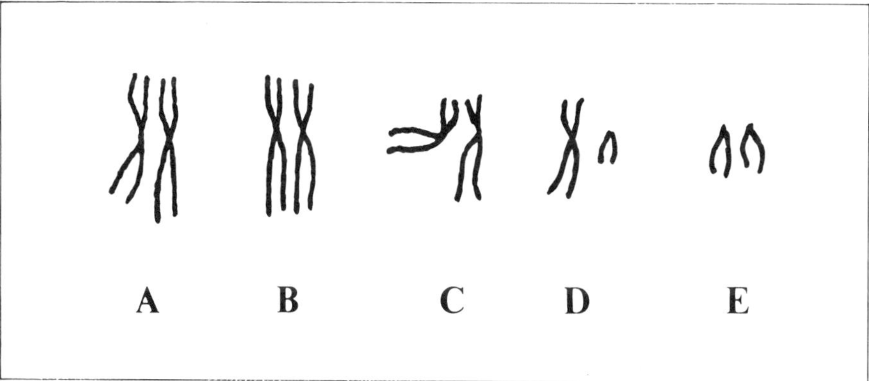

31 Which of the pairs of human chromosomes marked **A** to **E** decide which sex the baby will be?

32 How many chromosomes does a normal adult have?

A 22 **B** 23 **C** 44 **D** 46 **E** 47

33 How many chromosomes does the foetus get from the father's sperm? Use the list in question **32**.

34 How many chromosomes does a mongol have? Use the list in question **32**.

35 Which of these combinations of sex chromosomes will make the embryo a normal baby boy?

A XY **B** XX **C** YY **D** XXX **E** XXY

36 Which combination of chromosomes in the list above could *no* embryo have?

37 Where does the inherited part of your make-up come from? Does it come:

A mainly from your mother
B mainly from your father
C only from your mother
D only from your father
E equally from mother and father

38 Which of these hormones stimulates the growth of a new womb lining?

A oestrogen
B progesterone
C androgens
D thyroxine
E ACTH

39 Which of the hormones in the list above stimulates the adolescent growth spurt?

40 Which of the list is the so-called 'master hormone' from the pituitary gland, that speeds up the action of other hormones?

5. Senses

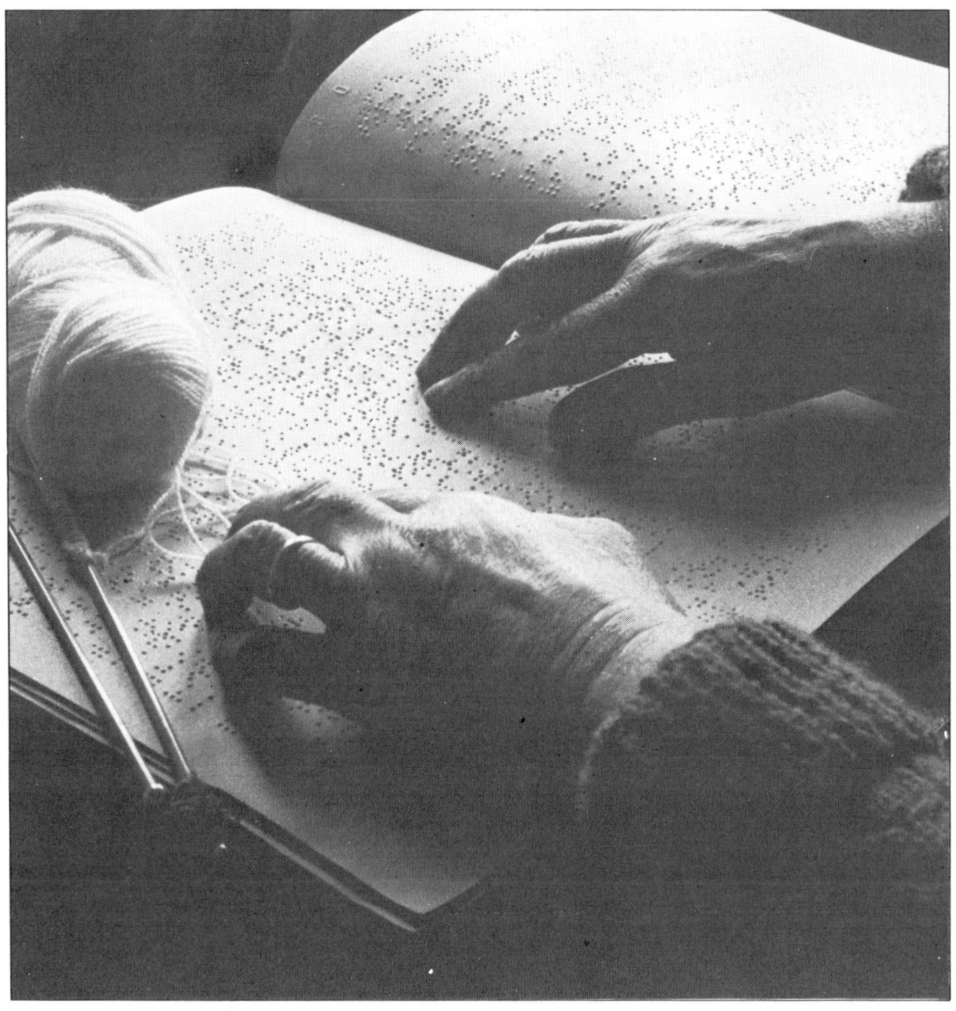

1 Our surroundings can change in a number of different ways. We call these changes *stimuli* and we can sense some of them. How many sorts of stimuli *can* we sense?

A 4 or less **B** 5 **C** 6 **D** 7 **E** 8 or more

2 We can sense *four* of the following stimuli. Which one are we *not* sensitive to? Is it change in:

A pressure **B** temperature **C** loudness
D brightness **E** radioactivity

3 Some instruments magnify a stimulus so we can sense it. Others change one we can never sense into one that we can. Which of the following is an example of the second kind?

A microscope **B** tape recorder **C** telescope
D radio set **E** stethoscope

4 Our eyes can make themselves less or more sensitive to one of the following stimuli.
Which one?

A brightness **B** pattern **C** colour
D movement **E** pressure

5 We have sensitive nerve endings buried in our muscles and tendons and ligaments.
What stimulus are *they* sensitive to? Is it change in:

A temperature **B** texture **C** weight
D pitch **E** chemical content

In the next five questions, choose the sense you would find most useful to do the job described. Use this list of senses:

A sight **B** hearing **C** touch
D taste **E** other senses

6 Finding your watch in a dark room.

7 Detecting a leaking gas pipe.

8 Deciding if the water in your bath is 'ready to get into'.

9 Finding out if your tea has been 'sugared'.

10 Deciding which of two sealed plastic containers has most flour in it?

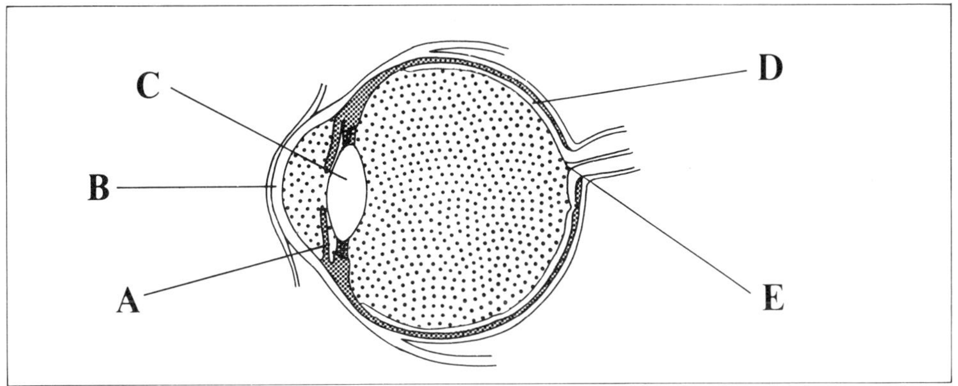

Choose one of the parts of the eye (labelled **A** to **E**) in the diagram, to fit each of the descriptions in the next four questions.

11 the part that is most sensitive to light

12 the part that adjusts the amount of light getting into the eye

13 the part that changes shape to focus the light

14 the part that gives your eyes their colour (e.g. '*blue* eyes')

15 You need *both* eyes working together to judge one of the following. Is it:

A distance	**B** brightness	**C** colour
D pattern	**E** movement	

16 Close your *right* eye and look at each of the numbers below. Start at the left and look at each number in turn. The cross should disappear when you are looking at a number near the middle.
What does this experiment show? Does it show that:

A your left eye is not as good as your right eye
B you need both eyes to see a cross
C after a while your eyes get tired
D you can only concentrate on one thing at a time
E you have a blind spot on your retina

17 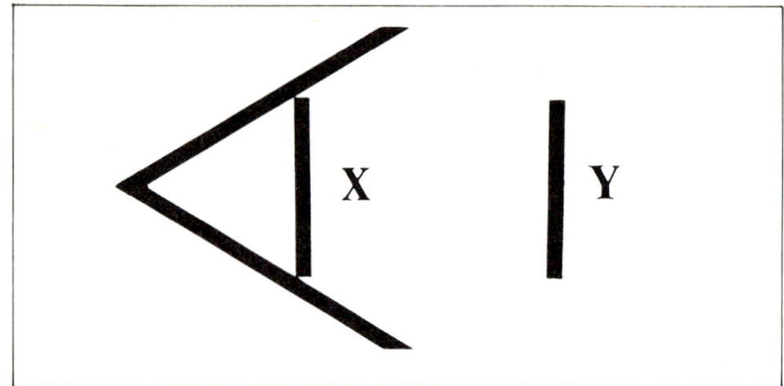

The two lines **X** and **Y** are the same length but one of them looks longer. What do observations like this tell us? Do they tell us that:

A seeing is believing
B artists can make mistakes
C you should not base judgements on single observations
D people cannot judge lengths very well
E upright lines should never be drawn near tilted ones

18 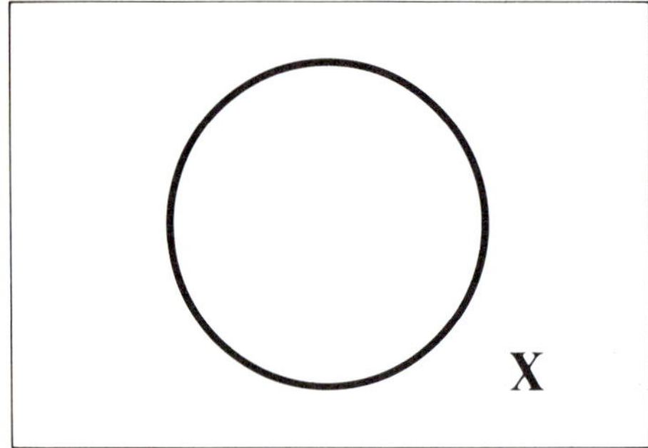

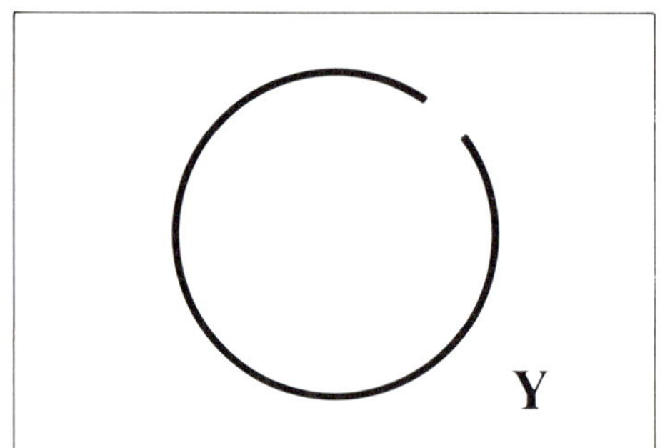

If the two cards, **X** and **Y**, were moved further and further away from you, a point would be reached where you could not tell that the **Y** circle had a gap in it. This method could be used to measure one of the following things to do with your eyes. Which one?

A visual acuity
B ability to focus
C ability to concentrate
D judgement of distance
E sensitivity

19 A cine film is a long strip of still pictures. Why then do the things in the pictures seem to be moving when it is projected? Is the best answer to this that:

A the film goes through the projector very quickly
B the film is moving instead of standing still
C film projectors use a special kind of light
D it takes time for the picture to die away on the screen
E the retina cannot follow fast changes in brightness

20 There is some truth in the story that vitamin A helps you see in the dark. Which of these statements to do with the story is/are true:

1 vitamin A helps your iris open wider
2 vitamin A helps to make visual purple
3 visual purple makes your retina more sensitive

Choose your answer letter using the code below:

Code	Choose					
		A	if	**1 2 3**	are all true
		B	if	**1 2**	only are true
		C	if	**2 3**	only are true
		D	if	**1**	only is true
		E	if	**3**	only is true

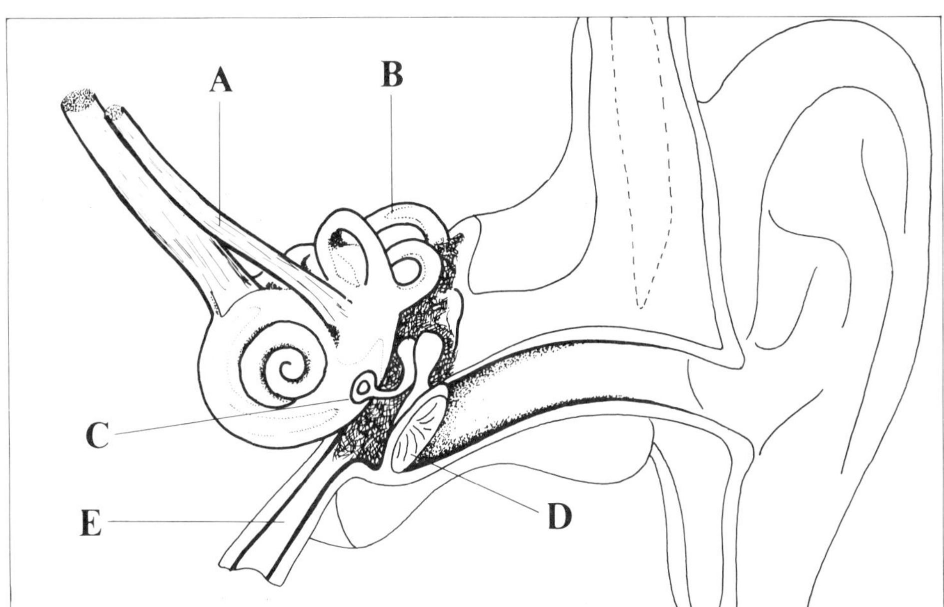

Choose one of the parts of the ear labelled **A** to **E** in the diagram to fit each of the descriptions in the next four questions.

21 the part that is known as the eardrum

22 the part that helps you to keep your balance

23 the part that balances the air pressure in the ear

24 the part that carries messages back to the brain

25 Some shepherds gives orders to their sheepdogs by blowing a special whistle. The dog can hear the whistle but the shepherd cannot. Is this because compared with us, dogs can:

A move their ears more
B hear softer sounds
C hear louder sounds
D hear higher pitched sounds
E hear lower pitched sounds

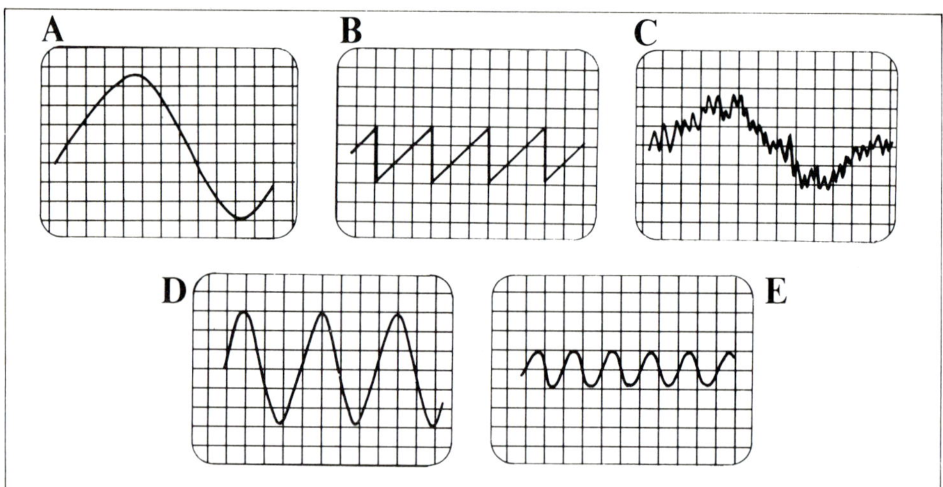

26 Different kinds of sound picked up by a microphone and fed into a cathode ray oscilloscope (CRO) produced the patterns shown in the diagram. Which sound (**A** to **E**) had the highest pitch?

27 Which of the patterns was produced by the loudest sound?

28 Which of the patterns was probably produced by somebody humming?

29 Which of the things in the list below changes electrical signals into sound vibrations?

A microphone
B loudspeaker
C amplifier
D record player stylus
E electric guitar string

30 Which of the things in the list above changes sound vibrations into electrical signals?

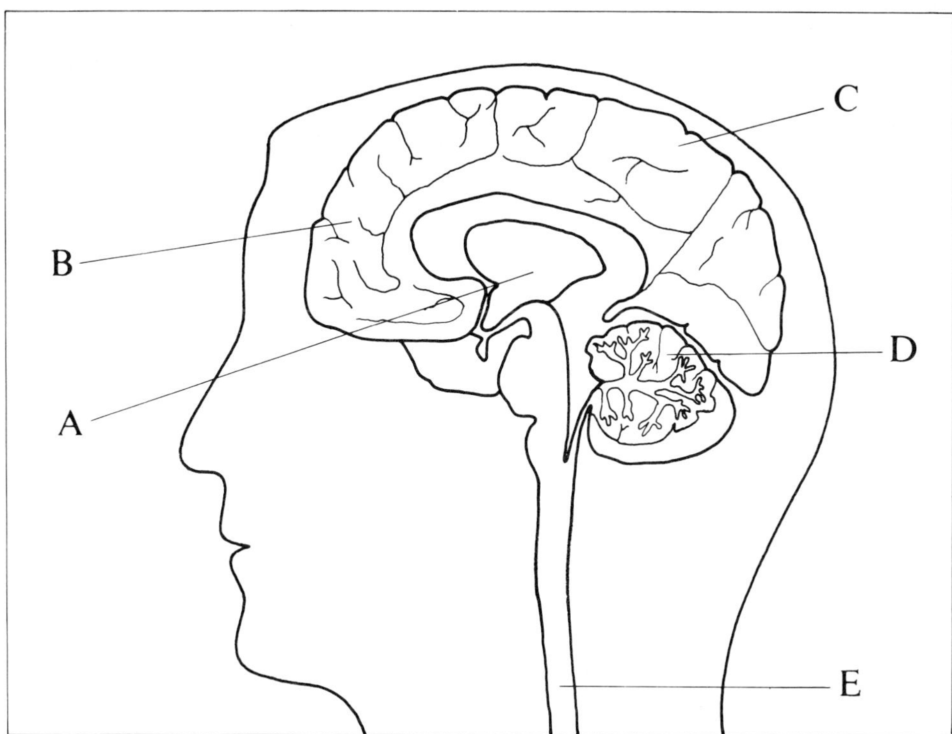

Choose one of the parts of the brain and spinal cord (marked **A** to **E** in the diagram) to fit each of the statements in the next five questions.

31 The place where sense messages are sorted out.

32 The part which has most to do with emotions and memory.

33 The place where the connections are made which produce reflex actions.

34 The part which has most to do with balance.

35 The part which sends out messages that make your muscles move.

36 The so-called 'autonomic' nervous system takes care of *four* of the following. Which does it *not* have much to do with?

A breathing
B walking
C balancing temperature
D digesting food
E keeping the heart going

37 Why are reflex actions (like pulling your hand away from something hot) so very fast? Is it because strong sense messages:

A travel faster through the nerve fibres
B go straight to the muscles
C are sorted out faster by the brain
D are passed on quickly by the spinal cord
E go through more nerve fibres

38 People working in kitchens learn to handle very hot things even though their reflexes try to make them drop them. Is this because the people:

A have thicker skins than usual
B can stand more pain than other people
C have got rid of their reflexes
D have destroyed the nerve endings in their skin
E use messages from the brain to stop their reflexes

39 The girl with the switch on the left is measuring the reaction time of the boy on the right. As soon as the light comes on he has to press a switch. The electric clock measures how long he takes. Which of the statements below is/are true?

1 the girl's switch stops the clock
2 the girl's switch starts the clock
3 the girl's switch puts the light on

Choose your answer letter using the code below:

Code	Choose					
	A	if	**1**	**2**	**3**	are all true
	B	if	**1**	**2**		only are true
	C	if		**2**	**3**	only are true
	D	if	**1**			only is true
	E	if			**3**	only is true

40 In an experiment like the one in question **39** the boy and girl drew a graph to show what happened to his reaction time as he had more and more tries. Which of the graphs below do you think they obtained?

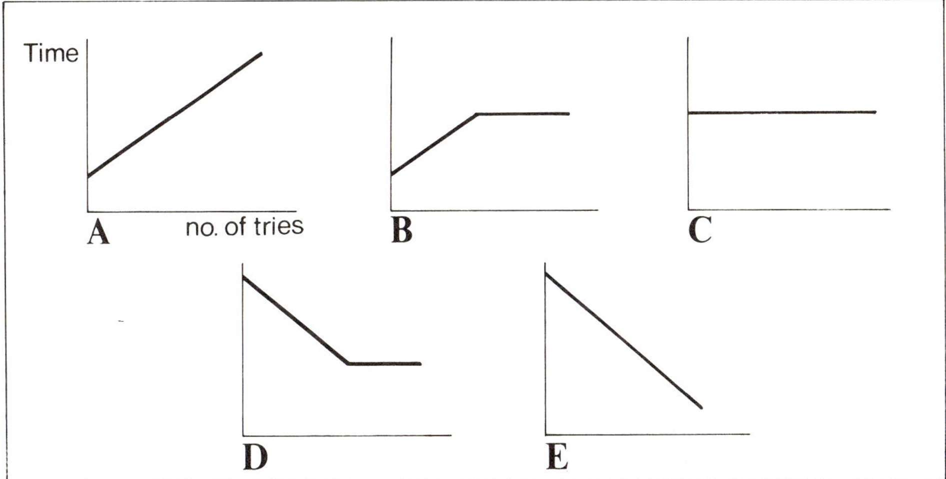